# Major Floods
# Of
# Pleasant Valley

On the cover: View from the east side of the bridge, looking into the village. The picture was taken near the crest of the 1955 flood. The picture was taken by H. Maynard Johnson.

# Major Floods
# Of
# Pleasant Valley

## By Pat Holt

Published by Burr Oak Publishing
www.BurrOakPublishing.com

Printed in the U.S.A.

# Acknowledgements

I would like to express my appreciation to the following individuals and organizations for their invaluable help in making this book possible: Marcia Johnson Valdata, Pat Silvernail Beatty, Jerry Ottaway, Caroline Dolfi, Edna Hommel, Terri Ghee of the Pleasant Valley Historical Society.

Unless otherwise noted, all images appear courtesy of the Pleasant Valley Historical Society, the photographer is unknown. Other images appear courtesy of H. Maynard Johnson (HMJ), Edna Hommel (EH).

# Pleasant Valley Flood of 1902

Although I could not find any newspaper articles or other information about floods other than the 1938 and 1955 floods, there were three pictures of a flood listed as 1902, among the other flood pictures in a folder, made by George Greenwood on the Historical Society computer.

This picture is listed as flood of 1902, North Avenue.

This picture is listed as flood of 1902, Quaker Lane and Main Street.

This picture is listed as flood of 1902, taken from South Avenue.

# Flood of 1938

September 1938, Pleasant Valley felt the effects of the wind and rain of what NOAA (National Oceanic and Atmospheric Administration) called "The Great New England Hurricane of 1938." (When hurricane naming practices began in 1953 hurricanes were named for women, prior to that they were named for the region affected.)

The combined effects from the recent system and the hurricane produced rainfall of 10 to 17 inches, resulting in some of the worst floods ever recorded in this area.

The Hurricane was also referred to by some sources as the "Long Island Express." According to NOAA, when it passed to the north of Puerto Rico on the 18th and 19th it was likely a category 5 hurricane. It turned northward on September 20th and by morning of the 21st it was 100 to 150 miles east of Cape Hatteras, North Carolina. At that point, the hurricane accelerated to a forward motion of 60 to 70 mph, making landfall over Long Island and Connecticut that afternoon as Category 3.

As the Wappingers Creek rose upstream from Pleasant Valley, two people, including a Pleasant Valley fireman, were drowned, in the raging waters near Lefty Atwater's Idlewild tavern on Creek road. The restaurant, about one mile northeast of the hamlet of Pleasant Valley, was near a small camp ground on the shore of the Wappingers Creek.

At least four other people, including members of the rescue party, were reported clinging to trees where they took refuge after being hurled into the water when their craft capsized. They had been in the trees from shortly before 10 o'clock the evening of September 21 to about 5 o'clock the following morning.

Kenneth Larkin, a 22 year old Pleasant Valley fireman and member of the first rescue party that rushed to a point opposite Lefty Atwater's tavern to take two women from a camp that was nearly submerged by water was reported one of the victims.

Dubois Haight, chief of the Pleasant Valley department, Chester Carlson and William Millard set out in a small boat and took the two women in the boat. They started for land when the boat upset, all five of them reached trees. Haight went back after those in the trees and took the two girls in the boat. The boat capsized and one of the girls was washed down the stream.

When the rescue boat overturned, Mrs. Keefe, 25, wife of Daniel J. Keefe, neighbor of Mrs. Ferguson at Astoria, took refuge in a tree for more than eight hours before she was finally rescued by Fireman Ralph Beacham of Arlington. Larkin was clinging to a tree when he attempted to reach Mrs. Keefe, and plunged into the rushing waters.

William Millard, it was said, gripped a tree near the spot where the boat overturned and clung there until 4:30 A.M. when a life line was struck to him and he was pulled to shore.

Displaying heroism in the battle to save those who had been hurled into the flood water, were Chester Carlson, DuBois Haight, chief to Pleasant Valley fire department, Ralph Beacham Arlington Fire department, Fire Chief Ghee, and Kenneth Martin, Arlington paid driver.

Members of the rescue party headed by Fire Chief Arthur L. Ghee of Arlington praised the courage of Mrs. Keefe, who secured her arms to a tree with vines and remained calm for more than eight hours until she was rescued.

The two victims, recovered from the muddy waters just south of the Idlewild tavern were identified as Kenneth Larkin, 22, son of Mrs. Florence Larkin, a Pleasant Valley fireman and employee of the Pleasant Valley Finishing company; and Caroline Ferguson, 21 of 330 49th street, Astoria, Long Island, New York.

Picture of Kenneth Larkin,
22 year old Pleasant Valley
Fireman
and hero who drowned while
attempting to rescue Ruth Keefe
from the raging waters of the
Wappingers Creek.


Photographer and date of picture
unknown.

    As the tragedy was taking place near Creek Road, downstream
in the hamlet water was rising and spilling over the banks of the creek
above the bridge and dam and flowing down Main Street into side
streets and homes, washing out yards and railroad tracks.

Gas Station, where town hall is now.

Pleasant Valley Ford Storage area, freight house on the left.
Facing toward Route 44 from rear of Ford Garage.

Pleasant Valley Ford Storage area, freight house on the left.

Washout on railroad tracks, looking west.

West Road at junction of Main St., Masten's Feed Store on right.
Presbyterian cemetary in the background across Main Street.

West Road, Dutchess Cider Corp.,later it was the saw mill, now is is
gone.

House behind Methodist Church.
The house is at the spot where the fire house is today.

Jordan's garage, Quaker Lane and Main St.
This was distroyed in the hurricane of 1955.

A back yard on South Avenue and the mill, from east bank of
Wappingers Creek.

*(Footnote from Poughkeepsie Eagle News article: Fireman Beacham
suffered from exposure yesterday morning and he went to Idlewild tavern
where he rested.)*

Endnote: During the years of prohibition the Lefty Atwater's establishment
was a speakeasy.

During the early years of television I remember my parents taking me to the
home of Lefty and Henrietta Atwater to watch TV on their small black and
white TV with an almost round screen.  The whole room was dark except for
the glowing images. We watched I Love Lucy and Lights Out Theater.

Flood 1902 and 1938 pictures are from a folder George Greenwood made of flood pictures on the Historical Society computer.

Photos property of Pleasant Valley Mill Site Museum, Photographers unknown.

1938 Flood information from the following sources:
Poughkeepsie Eagle News, September 22, 1938, page 1 & page 11
Poughkeepsie Eagle News, September 23, 1938, page 1 & page 2
http://www.erh.noaa.gov/er/nerfc/historical/sep1938.htm
http://www.erh.noaa.gov/box/hurricane/hurricane1938.shtml
http://www.nrc.noaa.govHAW2/english/history.shtml#new
http://www2.sunysuffolk.edu/mandias/38hurricane/
http://www.nhc.noaa.gov/HAW2/english/history.shtml

# Flood of 1955

August 1955 Hurricane Diane which followed close on the heels of Hurricane Connie, brought rain and wind and flooding to Pleasant Valley.

According to NOAA (National Oceanic and Atmospheric Administration) website, http://www.erh.noaa.gov/historical/aug1955.htm, Hurricane Connie produced generally 4 to 6 inches of rainfall on August 11 and 12, Hurricane Diane came a week later and the rainfall from Diane ranged up to nearly 20 inches over a two day period.

This was before the days of cell phones. Telephone Company operators in Poughkeepsie said there would be a delay of one and one-half hours in placing calls to Pleasant Valley, according to the Poughkeepsie New Yorker, page one article on August 19, 1955. It was also reported that there was no way to get into or out of Pleasant Valley. According to Town of Pleasant Valley Highway Superintendent Gleason approximately 500 families were affected by the flood after a day's rain which doused the area with seven inches of water.

Also reported on August 19, 1955 in the Poughkeepsie New Yorker:

*"A three room house completely furnished was swept from its foundation along the south side of Dutchess Turnpike just south of Pleasant Valley. The home was owned by Mr. and Mrs. William Hawley, and was occupied by Mr. and Mrs. William Houston. Mr. Houston is a photographer.*

*Mr. and Mrs. Houston evacuated their home during the night but 'When we left last night we didn't think it was so bad, so we didn't take anything but the clothes we wore.' Mrs. Houston said. 'We've lost everything.'*

*The house was seen about a quarter mile from its original site still being carried along by the creek waters."*

On the east edge of the hamlet where the town hall currently is, the water was said to be five to seven feet deep over the highway.

The paper further reported that the electric power had to be turned off to the town by Central Hudson Gas and Electric Corporation at 10:15 am.

Donald Robison, one rescue worker, was hauling stalled cars out of flooded areas when his tractor tipped over. He swam over to a nearby telephone pole which he climbed to await rescue.

According to the Highway superintendent, water in the village was between three and four feet high. It was up to the front of houses and stores and also inside some of the buildings.

Tractors and highway trucks were used to evacuate residents trapped in their homes by the flood. One resident, a teenager at the time remembers being taken in a dump truck from the center of the village to the Pleasant Valley School on Traver Road, where the Red Cross was to set up a shelter. She was worried because she had to leave her dog behind in her home. Later in the day the sun came out, she joined a friend to walk back to the village. They walked down Main Street where the water was knee high, not thinking about how contaminated the water might be. When she got to her home she found that her dog was fine.

According to the Poughkeepsie New Yorker August 20, 1955, Dr. Gordon C. MacKenzie, Town of Pleasant Valley health officer, warned residents whose homes were in the flooded area of the township, not to use water from their wells for drinking purposes; until tests were made by him and the State Department of Health Inspectors. Tank trucks with water were stationed in front of the firehouse, the school and the Grange hall.

A gas tank containing 1,800 pounds of bottled gas broke loose from the mill and was carried down the creek to Rochdale where it became wedged between two trees and a baseball backstop. (Almost 3 miles along the creek)

Reportedly eighteen caskets on show in the display room of the Allen funeral home in Main Street were ruined by the flood waters.

Several bottled gas and septic tanks snapped loose from buildings. Cows, horses and trees were seen being swept over the dam.

Highway superintendent Gleason said this flood was worse than 1938 in the amount of water and worse in damage, but this time it was better because the crest came at noon on a sunshiny day; the other time at midnight with a 50 mile wind blowing.

August 20, 1955 Poughkeepsie New Yorker reported:

*"Yesterday's floods were worse than the 1938 hurricane, according to Professor A. Scott Warthin, geologist at Vassar College and chairman of the Dutchess County Water Conservation committee.*

*While the torrential waters in Wappingers creek washed away the equipment of the U. S. Geological Survey station there. Professor Warthin estimated the peak flow at 17,000 cubic feet per second. That flow compared with a peak of 15,900 cubic feet per second reached during the 1938 hurricane. Yesterday's peak at Red Oaks Mill occurred at 3:30 p.m."*

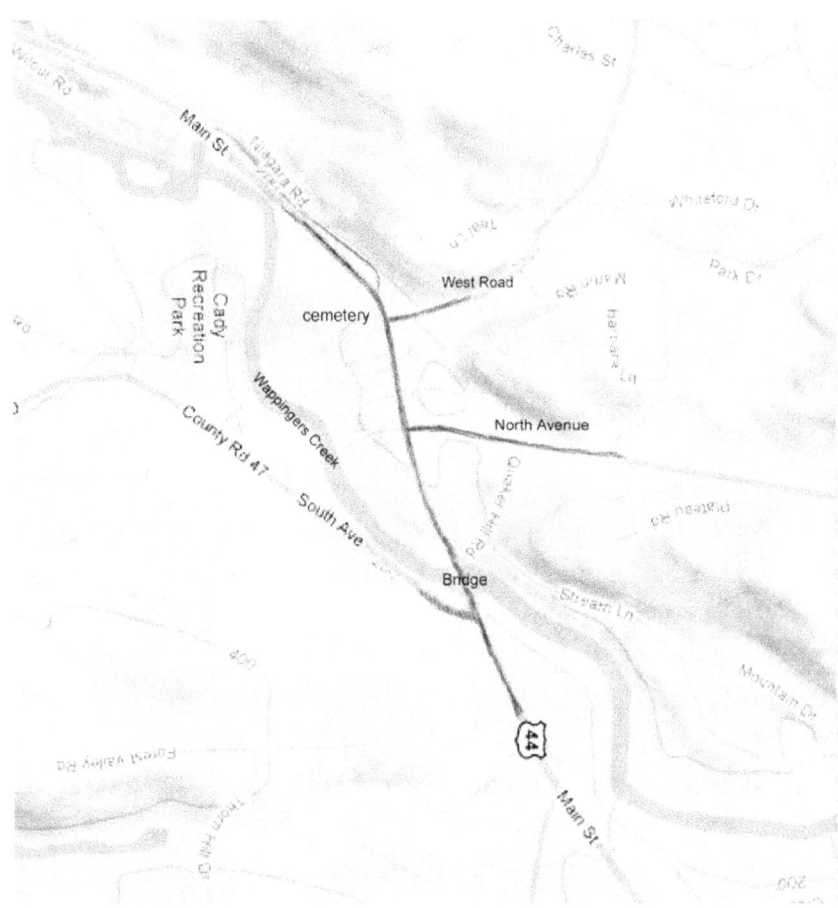

This map from map.google.com shows the path of the Wappingers Creek as it flows through the village of Pleasant Valley.

The Cady Recreation Park, view from Route 44, water from the lower portion of the Wappingers Creek is covering the, ball field, picnic tables, and swing set.

As the lower portion of the Wappingers Creek backed up and overflowed its banks, the water covered portions of Route 44, and was moving through the cemetery, onto Main Street. In some of these pictures cars are still able to drive through the water.

Following are some pictures taken at various stages of the flood.

H. Maynard Johnson lived near the Ford Garage, and took many pictures of the scene as the water became higher.

The water coming from the Wappingers Creek near the bridge flowing down Main Street joined with the water moving up Main Street, and the entire street including the houses and business in the village were flooded.

Local highway department employees and volunteers from the fire department, and anyone with a big truck or tractor braved the flood waters to rescue residents of Pleasant Valley.

The Texaco station shown here, near the junction of West Road and
Route 44 or Main Street is now a bicycle shop.(HMJ)

Don Robison driving his tractor on the lookout for people to
rescue.(HMJ)

The water from both ends of town have met and flooded Main Street.(HMJ)

Town truck rescuing stranded residents. (HMJ)

The Town truck with town employees on the job.(HMJ)

The gas pumps shown here have been replaced by the CVS driveway.(HMJ)

Looking west on Main Street. (HMJ)

LLooking west on Main Street. (HMJ)

Gas pumps in front of the Pleasant Valley Ford Garage, now CVS.(HMJ)

Men looking at the water from the cemetery wall. (HMJ)

Water flowing over the stone wall fence surrounding cemetery. (HMJ)

The cemetery (HMJ)

View of the cemetery looking from Main Street toward the Wappingers Creek.(HMJ)

The Ford Garage which has been replaced by CVS.(HMJ)

Thorne Johnson and Babs Lovelace resting on the cemetery wall.
(HMJ)

Picture taken from a tractor.

The Presbyterian Cemetery after the waters receded.

The Presbyterian Cemetery after the waters receded.

Mrs. Lena Johnson, Marcia Johnson holding baby Steven Nosonowitz, Mrs. Nosonowitz, Anna Millard at the Traver Road School waiting for the Red Cross. (HMJ)

Car leaving Pleasant Valley going toward Poughkeepsie on Main Street.(HMJ)

Car coming out of West Road onto Main Street. (HMJ)

The building in the center of this picture is the G. E. Masten Feed Store, run by Donald and Wesley Drake. It is still the G. E. Masten Feed Store, but now it is being run by Pam and Don Cady. The Sunoco gas pumps and the Ford Garage are gone and replaced by the CVS Pharmacy and parking lot.

Main Street, early in the flood. (HMJ)

A crowd is gathering. (HMJ)

People seem to like standing and walking in the flood waters. (HMJ)

Looking across Main Street as the water rises. (HMJ)

This could have been near the crest. (HMJ)

The G. E. Masten Feed Store from the cemetery. (HMJ)

The front view of the feed store. (HMJ)

Feed store from near the gas pumps at the Ford Garage. (HMJ)

Men leaving the Feed Store, it looks like it could be
owners, Donald and Wesley Drake.

Pleasant Valley Garage, known locally as "The Ford Garage" (HMJ)

Currently this is the site of the CVS Pharmacy.
(HMJ)

Looking across Main Street toward West Road. (HMJ)

Looking up West Road from Main Street.(HMJ)

Looking up West Road. (HMJ)

Looking up West Road from Main Street, note the debris around the
road sign near the center of the photo.(HMJ)

The house in the center is where Wesley and Donald Drake lived.
(HMJ)

People walking and standing in the flood water. (HMJ)

The rear of the Ford Garage, looking toward West Road, note the floating milk can.(HMJ)

(HMJ)

Pictures taken from the back of the Ford Garage.(HMJ)

A sunken suburban behind the garage. (HMJ)

As Mr. Johnson started to look up Main Street toward the village, these are the pictures he took.

The Mid-Way Burner Parts Co. is no longer there, the building is gone, it is part of the parking lot for the dentist office next door. (HMJ)

The Presbyterian Church is on the right, and up enough of a hill that it didn't flood. (HMJ)

The Department Store is still there, the tree in front of it is gone. (HMJ)

This picture was most likely taken from a tractor during the crest of the flood.

Looking West on Main Street from a truck near the
Department Store. (HMJ)

Looking east on Main Street, sign for Talbot's on the left. (HMJ)

Talbot's Inn, later was known as the 1830 Inn, at the northwest corner of Main Street and North Avenue. This historic structure was razed to make way for a shopping mall.

South side of Main Street, looking from North Avenue across, toward the Pleasant Valley Hotel. This historic structure was razed to make way for a shopping mall.

View of the corner of the Pleasant Valley Hotel, the Pleasant Valley
hotel was also known as Cady's, since it was owned by Mike Cady,
long time fire chief of the Pleasant Valley Volunteer Fire Department.

Looking at the center of the Village. (HMJ)

Looking from North Avenue toward small strip mall along Main Street.
The strip mall in this picture was torn down to make way for a small
shopping center in about 1971. (HMJ)

Elmer Nygren's GE Appliance Store.

This was Conklin Factory; it was near the creek, behind the small Strip mall. The factory has since moved to West Road.

View from tractor, first looking toward village from near the Pleasant Valley Department Store, second looking away from village toward the Presbyterian Church and cemetery from the same spot.

Main Street center of village. (HMJ)

A little farther up Main Street. (HMJ)

Main Street, unknown people I the middle of the street. They
could have been watching the Post Master and postal workers
carry mail from the Post Office.

Post Master "Pat" Clark with Edna Hommel, Chub Parks, Dorcus Brower and
Mary Parks carrying the mail out of the flooded Post Office.(EH)

Ed Smith driving the truck, Chub Parks and Post Master "Pat" Clark
carrying boxes of mail, Edna Hommel on the truck.(EH)

Mary Parks and Post Master "Pat" Clark outside the back door
of the Post Office.(EH)

Mary Parks, Edna Hommel, Dorcas Brower, "Pat"Clark
Inside the post office. (EH)

Mary Parks inside the Post Office. (EH)

Postal workers in Main Street, near the front of the Post Office, with the mill water tower in the background.(EH)

Main Street looking west area in front of the firehouse.

Looking at the fire house from across the Main Street.

The dog on the porch is watching the flood waters rising. (HMJ)

    This house was next door to the Methodist Church on Main Street; James Kepfort owned the house and rented it as two apartments. It had previously been the home and office of a dentist, Dr. Tinkleman.
    Both of the Methodist Church and this house are gone now, and replaced by the driveway to the firehouse and Jack Haverty's Napa Store.

Main Street, Sergeant Howard Moore of the State Police, who lives near where picture was taken, is talking to a town worker. (HMJ)

Workers from the Pleasant Valley Finishing Company are out on the sidewalk. (HMJ)

Main Street, looking toward the Pleasant Valley Grill, Quaker Hill Road and the Potash house. It looks like these men are starting to do some clean-up.

The scenes along Main Street show water in the street, filling the stores and basements, or covering the cemetery. They do not show the real power of the water. As you get closer to the Wappingers Creek, the real power of the raging flood water is apparent.

Looking at Quaker Hill Road, during the time the flood was at its worst, you could see the water came around the side of the bridge, through the large house on Quaker Hill Road and destroy this garage. Pictures from the Pleasant Valley Historical Society list the name of the building as "Jordan's Garage"

This story was reported in the August 20 Poughkeepsie New Yorker:
*"A large barn that was attached to the rear of the building occupied by the Pleasant Valley grill was washed away. It floated about 50 feet from its site and pounded against a tree in Quaker Hill road.*

*The remains of the barn looked like an old covered bridge. The sides were torn away by the wild waters."*

The following four pictures taken from different angles show the scene as the garage is being destroyed by the power of the water:

Jordan's Garage on Quaker Hill Road, the water tower of the mill can be seen in the background.

Looking toward Main Street and the bridge from Quaker Hill Road.

Looking toward Quaker Hill Road at what is left of Jordan's Garage.
(HMJ)

Corner of Main Street and Quaker Hill Road, looking at Potash house, which is currently Fred Schaefer's Pleasant Valley Office.(HMJ)

At one point the water was going in the back windows of this house and out the front. At the time this house belonged to Brinkerhoff.

Brinkerhoff's house, picture taken from the bridge. This house is now divided into apartments and belongs to Craig Domonkos. (HMJ)

The crowd is gathering on the bridge.(HMJ)

Looking from east side of the bridge toward the town. This must have been near the crest, since the water is touching the bridge. (HMJ)

Another view of the Wappingers Creek overflowing into town. (HMJ)

Looking downstream from the bridge, the ripple in the water is where the six foot high dam was. On the right is the Pleasant Valley Finishing Company. (HMJ)

Pictures taken from the bridge looking downstream.

View from the bridge over the Wappingers Creek, looking north. They might be pointing at the cows or horses that were reported to have been carried away by the raging water.

Looking up toward the bridge from a backyard on South Avenue.

Looking south on North Avenue, toward Main Street, from near Quaker Hill Road.

House near the firehouse, on the left, both are gone now, the firehouse in this picture was razed to make way for a new bigger one.

One of the larger bridges in the county, a 100 foot long span that crossed the Wappingers Creek on Hurley Road, near Salt Point was destroyed.
(HMJ)

Where the bridge was connected (HMJ)

The skeleton of the bridge lying in the Wappingers Creek. This bridge
survived the great flood of 1938. (HMJ)

The skeleton of the bridge lying in the Wappingers Creek. This bridge survived the great flood of 1938. (HMJ)

Picture of a house what was swept off its foundation. This house was near the Hurley Road bridge. (HMJ)

Picture of a house near Hurley Road Bridge that was swept off its
foundation (HMJ)

Pelham's house and garden, after the flood. The property is west of
Pleasant Valley along Route 44, now owned by "The Cutter's Edge" hair
salon. (HMJ)

Pelham's house and garden, after the flood. (HMJ)

Either's foundation and house after the flood, Either's house was near Pelham's.(HMJ)

Either's house, swept from its foundation. (HMJ)

Flooded field, unknown location.

Route 44, east of the village of Pleasant Valley near Albrect's farm.
(HMJ)

Route 44, east of the village of Pleasant Valley near Albrect's farm.
(HMJ)

During the early 1960's the Army Corps of Engineers lowered the Pleasant Valley dam by three feet and built a retaining wall on the southwest side of the bridge. Since then the Wappingers Creek has not overflowed in that area.

The Creek still backs up on the west side of town and floods the Cady Recreation Park.

Endnotes for 1955 flood:

http://www.erh.noaa.gov/historical/aug1955.htm
Poughkeepsie New Yorker, page one article on August 19, 1955
Poughkeepsie New Yorker August 20, 1955

1955 flood pictures are from a folder George Greenwood made on the Historical Society computer and other more recently donated pictures.

Photos property of Pleasant Valley Mill Site Museum, photographers unknown, unless otherwise indicated.

Known photographers:
H. Maynard Johnson – HMJ
Edna Hommel - EH